curious?

SAVING WATER

BY AMY S. HANSEN

AMICUS LEARNING

What are you

curious about?

CHAPTER THREE

Take Action

PAGE
16

Curious About is published
by Amicus Learning,
an imprint of Amicus.
P.O. Box 227
Mankato, MN 56002
www.amicuspublishing.us

Editor: Alissa Thielges
Series Designer: Kathleen Petelinsek
Book Designer and Photo Researcher: Emily Dietz

Library of Congress Cataloging-in-Publication Data
Names: Hansen, Amy, author.
Title: Curious about saving water /
by Amy S. Hansen. Other titles: Saving water
Description: Mankato, MN: Amicus Learning, [2025] |
Series: Curious about green living | Includes bibliographical
references and index. | Audience: Ages 5–9 | Audience:
Grades 2–3 | Summary: "Ignite kids' growing curiosity
about the environment with an inquiry-based approach
to water conservation. Includes infographics and back
matter to support research skills, plus table of contents,
glossary, and index"—Provided by publisher.
Identifiers: LCCN 2023043280 (print) | LCCN
2023043281 (ebook) | ISBN 9781645496991
(library binding) | ISBN 9781681529684
(paperback) | ISBN 9781645497059 (ebook)
Subjects: LCSH: Water conservation—
Juvenile literature. | Water—Juvenile literature.
Classification: LCC TD388 .H365 2025 (print) | LCC
TD388 (ebook) | DDC 333.91/16—dc23/eng/20231107
LC record available at https://lccn.loc.gov/2023043280
LC ebook record available at https://lccn.loc.gov/2023043281

Photo Credits: 123RF: nikkikii, cover, 1; Alamy: SOPA
Images Limited, 2, 8–9 (l); Getty: feellife, 13 (1), Goodboy
Picture Company, 18, Joa_Souza, 5, Mindful Media, 21,
sharply_done, 13 (3), Unya-MT, 3, 20, Willowpix, 6; Noun
Project: Monika, 22 & 23 (water), Adrien Coquet, 22 & 23
(river); Shutterstock: Albina Bugarcheva, 13 (2), azure1, 13 (5),
Bilanol, 14–15, Evgeny Haritonov, 16–17, Labellepatine, 13 (4),
Marco Tulio, 7, New Africa, 11, pikselstock, 19, sima, 2, 12 (r)

Printed in China

CHAPTER ONE

Why should I save water?

Most of the Earth is covered in water. It seems hard to worry when there is so much water. But ocean water is salty. A lot of water is **polluted**, or dirty. Earth has only a tiny bit of clean, **fresh water**. We all need to drink fresh water to stay healthy.

AMOUNT OF WATER ON EARTH

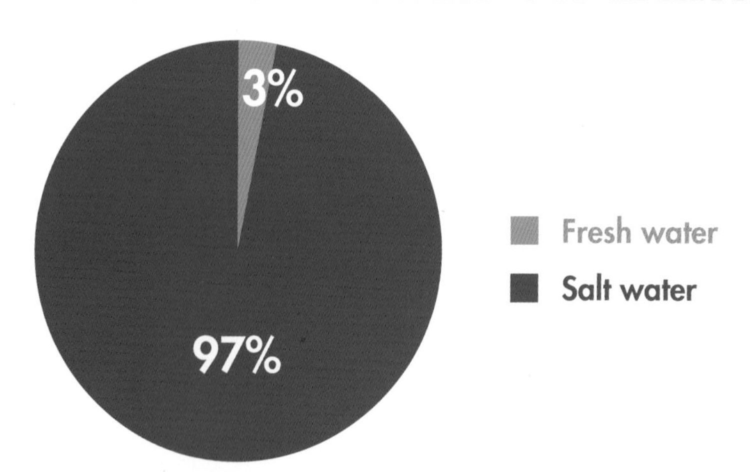

Litter pollutes oceans, lakes, and rivers.

Plants need rain to live, but too much creates a flood.

Does water get lost in the water cycle?

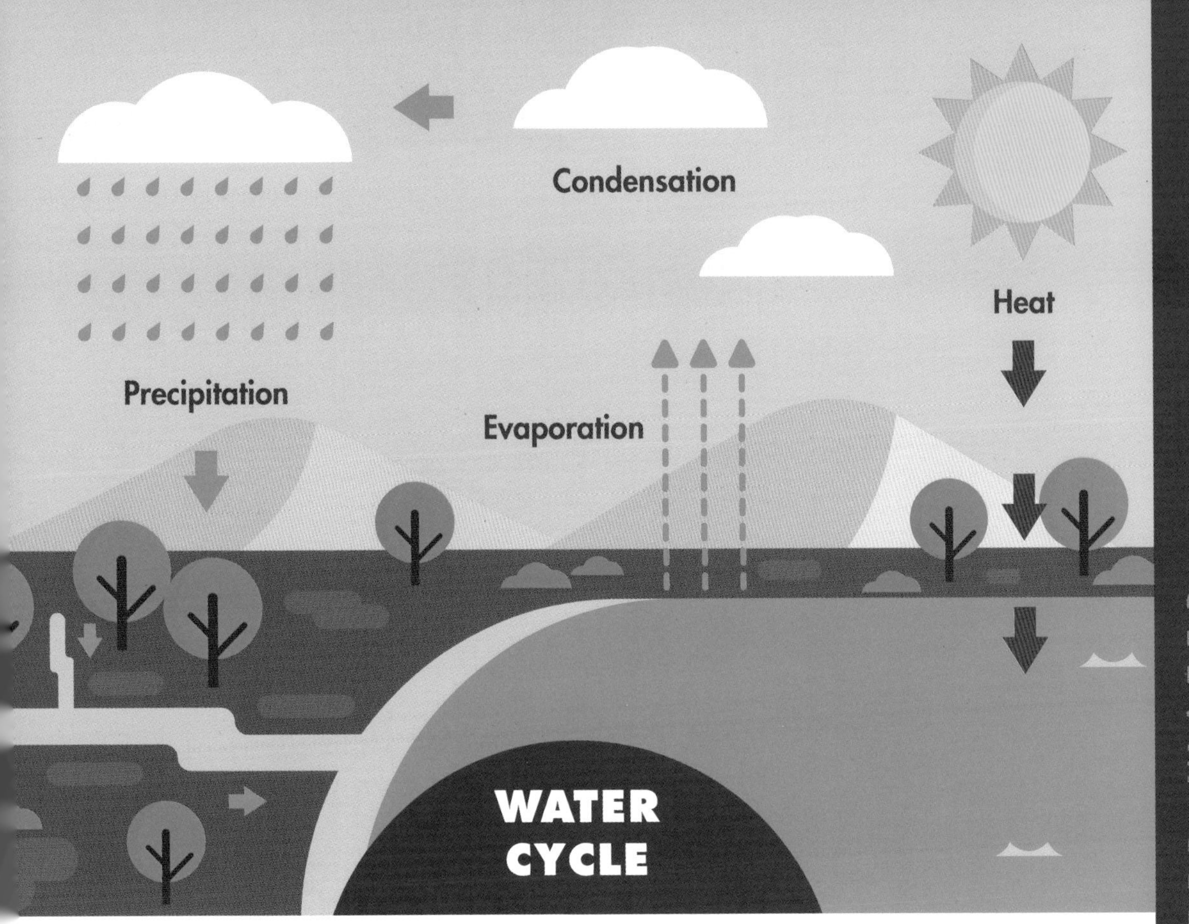

No. Earth never gains nor loses any water. But the **water cycle** covers the whole planet. When a puddle dries by your house, the water travels through the air. It comes down as rain somewhere else. The water isn't lost, but there may not be any more water nearby. Some places get too much rain and the area floods. Other places get too little. They may be in a **drought**.

Is the Earth running out of water?

A lake in Nevada dried up in 2021 due to a major drought.

DID YOU KNOW?
When it is dry, the ground is too hard. It can't soak up water and rain runs off. This is one of the reasons flooding happens after a drought.

Some places are. In the United States, the southwest has been in a drought for years. That area gets its water from rivers. Rivers carry water from far away. But they are running dry, too. This hurts people who live where water used to be. It also hurts the plants and animals that need the water. We all need water to live.

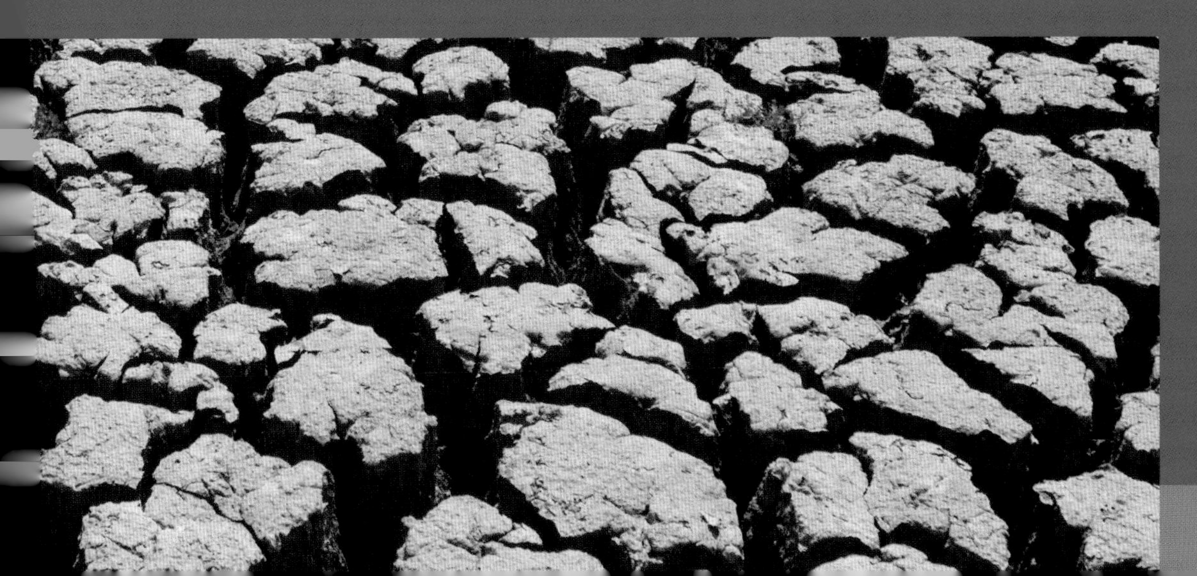

CHAPTER TWO

How do we all use water every day?

We drink water. And we drink juice that has water. We use the toilet. We take showers or baths. We do laundry. We make dinner and wash dishes. All of these things use water. Most families use about 300 gallons (1,136 liters) of water a day.

Handwashing dishes can use more water than a dishwasher.

Who uses the most water?

Large sprayers water a field of corn.

Farming uses the most. Farmers spray water on their crops. Farm animals also drink a lot of water. Many other jobs and **factories** need water, too. Power plants use it to make electricity. And clothing makers use water. They need it to dye the cloth.

1

AGRICULTURE
WATERING PLANTS

2

CLOTHING
DYEING CLOTHES

3

POWER PLANTS
USING STEAM TO MAKE ELECTRICITY

4

MEAT INDUSTRY
GIVING WATER TO ANIMALS THAT WE USE FOR FOOD

5

BEVERAGE INDUSTRY
MAKING THE LIQUIDS AND PACKAGING

A treatment plant removes harmful chemicals, solids, and gases from water.

Can we clean dirty water?

DID YOU KNOW?
Scientists can take salt out of ocean water. But it takes a lot of energy.

Yes! Dirty water goes to a treatment plant. Machines send the water through filters. They also treat the water with chemicals to kill germs. When they are done, the clean water goes back into a nearby body of water.

Is anyone working to save water?

Yes! Scientists are helping rivers. They talk with farmers. Farmers figure out how to use less water. Then rivers have more water. Also, **engineers** build ponds to hold water. Rainwater moves through storm drains. Water slows down in the pond. Plants and animals can drink that water.

A researcher tests water and soil to see if they are polluted.

What can I do to save water at home?

Keep the faucet off when you are brushing your teeth to save on water.

Turn off the faucet quickly. If you see a leak, tell an adult. Take short showers. And turn the water off when you brush your teeth. Use a barrel to catch rainwater from your roof. Then you can use the water for a garden.

DID YOU KNOW?
You can use up to 27 gallons (102 L) of water to wash dishes by hand. A good dishwasher uses only 3 gallons (11 L) per load.

Kids help take water samples for testing.

How can I help other people save water?

With an adult, join projects that protect water. Some **volunteers** do water testing. Others raise fish to release into the rivers. Other projects plant trees along riverbanks. Make sure to visit wild rivers, too. You can see how nature uses water. It will remind you why saving water is important for all.

DID YOU KNOW?

Planting native plants in your yard saves water. These plants are used to the amount of rainfall your area gets each year.

Planting native flowers will help your garden last longer.

ASK MORE QUESTIONS

Why is the ocean salty?

How important is groundwater?

Try a BIG QUESTION: How can laws and governments help save water?

SEARCH FOR ANSWERS

Search the library catalog or the Internet.
A librarian, teacher, or parent can help you.

Using Keywords
Find the looking glass.

Keywords are the most important words in your question.

?

If you want to know about:

- the ocean's salt content, type: OCEAN SALINITY
- how important groundwater is, type: GROUNDWATER IMPORTANCE

FIND GOOD SOURCES

Are the sources good?
Some are better than others. An adult can help you. Here are some good, safe sources.

Books

Clean Water for All
by Danielle Haynes, 2022.

Conserve It! by Mary Boone, 2020.

Internet Sites

Water Use It Wisely
https://wateruseitwisely.com/kids-teachers/fun-activities/
This program spreads awareness about saving water. Its site has resources and games for kids.

Let's Make a Change: Water Pollution!
https://natgeokids.com/uk/hidden-category/lets-make-a-change-water-pollution/
National Geographic explores the planet. It is a good source about animals and nature.

Every effort has been made to ensure that these websites are appropriate for children. However, because of the nature of the Internet, it is impossible to guarantee that these sites will remain active indefinitely or that their contents will not be altered.

SHARE AND TAKE ACTION

Find out where your home water comes from.
Ask an adult to help you research.

Take a colder shower.
Jump in right away so you don't waste water waiting for it to warm up. Or catch the water that comes out. Can you use it somewhere else?

Talk with family about saving water.
Look up tips online. See how much water you can save.

GLOSSARY

drought A long period of time during which there is very little or no rain.

engineer A person who has scientific training to design and build complicated products, machines, systems, or structures.

factory A place where materials are put together, usually with machines.

filter A device that is used to remove something unwanted from a liquid or gas that passes through it.

fresh water Relating to water that is not salty.

native Belonging to a particular place.

pollute To spoil with waste made by humans.

volunteer Someone who does something without being forced to do it.

water cycle The continuous movement of water between lakes, rivers, oceans, the atmosphere, and land in liquid, gas, and solid states.

INDEX

About the Author

Amy S. Hansen lives in Maryland. She started writing as a reporter for local newspapers. After a few years she went back to school to study environmental science and then worked as a science reporter. Nowadays, she (mostly) writes for kids because it is much more fun.